Einstein

D0482910

The Impossible Shrinking Machine

and Other Cases

Seymour Simon
Illustrations by Kevin O'Malley

Volume **1** in the Einstein Anderson Series

 **SEYMOUR SCIENCE**

Published by Seymour Science LLC.

These stories, which have been substantially updated and expanded for a new audience, are based on the Einstein Anderson book originally published in 1980 by Viking Penguin, New York, under the title *Einstein Anderson, Science Sleuth*, and republished in 1997 by Morrow Junior Books, New York, under the title *The Howling Dog and Other Cases*.

Contact: Seymour Science LLC,
15 Cutter Mill Road, Suite 242,
Great Neck, NY 11021.
www.SeymourSimon.com

www.StarWalkKids.com

ISBN: 978-1-936503-05-6

Contents

The Extreme Rollerblades

It was the best day of the year. At least that's what Einstein Anderson thought.

School had let out in the town of Sparta and today was the beginning of summer vacation. Two months of sleeping late in the morning. Two months of doing what Einstein wanted to do, not what Ms. Sugar, his fifth-grade teacher last year, wanted him to do.

Einstein knew it was the first day of vacation, but his body thought it was time to go to school. Even though his alarm hadn't gone off, he'd woken up anyway. He lay awake in bed, with sunlight streaming in under the curtains, then he rolled over to check the time on his phone. It was 7:15 in the morning. Einstein Anderson pushed his head deeper into his pillow. He guessed he would get up in another hour or so. That is, if he felt like getting up.

Einstein's real name was Adam. But nobody called him Adam anymore, not even his parents. Einstein had been interested in science for as long as he could remember. From an early age he had solved science problem after science problem that stumped even his teachers.

At the age of six Einstein explained to Ms.

Moore, his kindergarten teacher, how to make a real diving submarine out of a tall jar, a medicine dropper, a rubber band, and a balloon. At the age of seven Einstein showed Ms. Patrick, his first-grade teacher, how to set up a balanced aquarium in the classroom. At the age of eight Einstein built a robot that won first prize at the statewide science fair.

It was Ms. Moore who gave Adam the nickname of Einstein. Soon all his friends called him Einstein. Adam was proud of his nickname. He knew that Albert Einstein was the most famous scientist of the twentieth century, who had discovered many important things about the universe. His equation $E=mc^2$ led to the understanding of atomic energy. Albert Einstein had been a gentle, kind man as well as a genius.

But even Albert Einstein enjoyed his summer vacations.

Twelve-year-old Einstein Anderson closed his eyes and tried to drift off back to sleep. He must have succeeded because the next thing he noticed was the theme song from *Star Wars*. It was his phone alarm. With a groan he rolled over, put on his glasses, and looked at the screen. It was now 8:15 and he had a text from his friend Paloma.

It read, **"Check out Stanley's blog!!!"**

Einstein groaned again. Stanley Roberts was a kid in Einstein's grade who always had a scheme to make money. He saw himself as the next Steve Jobs or Mark Zuckerberg, someone who would use science and technology to become a billionaire. The problem was that Stanley didn't ever bother to learn anything about science. His money-making schemes always turned out to be fakes, one way or another.

Still in his pajamas, Einstein sat down at the

small desk in his room and turned on his laptop. In a few seconds he was looking at the blog page for StanTastic Industries, Stanley's made-up tech company.

"Technological breakthrough!" the page said in blinking letters. **"Greatest Rollerblades ever! X-treme! New technology makes you fast, fast, fast! Going on sale soon—reserve yours now!"**

Einstein knew that even though it was the first day of vacation, he had work to do. Most kids at school knew enough not to hand over money to Stanley, but he might be able to fool some of the new kids. Picking up his phone he thumbed in a reply to Paloma.

"Let's C Stan this AM"

He'd have to find out more about this Rollerblade scheme of Stanley's, but first he needed breakfast. He went into the bathroom, washed, and got dressed in T-shirt, sneakers, and his favorite weekend jeans, broken-in and ragged at the knees.

Einstein was an average-size twelve-year-old boy. His light brown eyes were a little nearsighted, and his glasses seemed a bit too big for his face. His eyes sometimes had a far-away look, as if he were thinking about some important problem in science. But Einstein was not always serious. He loved a good joke (or even a bad one) and liked to make puns, the worse the better.

Matt Anderson, Einstein's father, looked surprised when his son walked into the kitchen and sat down at the table. He was just finishing breakfast and was about to leave for his office. Einstein's dad was a veterinarian. His little brother Dennis had his face buried in a plate of pancakes.

"To what do we owe the honor of your presence so early in the morning, Einstein?"

asked his father with a smile. "I was sure you were going to sleep in on your first day of vacation."

"I got a text from Paloma," said Einstein. "Stanley has a new plan—some kind of extreme Rollerblades."

"I guess science never takes a day off," his father said with mock sympathy.

Einstein sniffed the air. "Any pancakes left, Mom?" he asked hopefully. "I just have time for a light breakfast."

Emily Anderson, Einstein's mother, nodded from her spot by the stove and watched with amusement as her son walked over to the refrigerator, poured himself a glassful of orange juice, popped two slices of bread into the toaster, poured himself a full bowl of milk and breakfast cereal, and sat down to eat.

His mom laughed, "Well I'm glad you have time for a light breakfast. I'd hate to see what a heavy one looked like." While pancakes sizzled, Emily Anderson turned to a laptop she had perched on the kitchen counter and typed in a few words. She worked as a writer and editor on the *Sparta Tribune*, the town's paper.

"That reminds me," Einstein said with his mouth full of cereal. "What do pancakes and baseball teams have in common?"

"I don't know," his mom said with a groan. "But I bet you do."

"Batter, batter, batter!"

"I don't get it," said Dennis, who was eight. He swallowed a last bite and jumped up, pushing his chair back with a loud squeak.

"Ow!" Dennis cried. "What a noisy chair."

"The chair's not noisy," Einstein said, biting into a piece of toast. "The sound is from the friction of the chair legs against the floor."

"Fiction?" Dennis asked. "You mean like,

science fiction?"

"No, not fiction," Einstein explained, trying to be patient. "*Fric*-tion." He often thought Dennis got words wrong just to annoy him. "Friction happens whenever two objects rub against each other. It's the force that slows them down or stops them. The noise was from the energy released by the friction of the chair legs dragging along the floor."

"Sounds like fiction to me," Dennis said. Einstein didn't bother to respond.

"Hurry up, Dennis," said Emily Anderson. "Carl's mom should be picking you up for day camp any minute now."

Just as she said it, a car horn outside beeped loudly. "And don't forget your lunch!" she shouted, as Dennis headed out the door. Dennis came back, took the bag lunch from his mom, and suffered through a kiss on the forehead. Then, sticking his tongue out at Einstein, he ran out the door.

"See you!" he shouted as the door slammed shut.

Matt Anderson got up. "Well, time to go help some sick pets," he said. He kissed his wife good-bye and ruffled the hair on Einstein's head.

"There's some friction for you," he said with a laugh, as he went out the door.

"I've got to get going, too," Emily said, looking at the clock. "Will you be all right till I get home this afternoon? Soccer camp doesn't start until next week."

Einstein wolfed down his breakfast and put his plate in the sink.

"Sure Mom, I'm twelve now," he assured her. "Besides, Paloma and I have a case to work on."

"Well, don't be too hard on Stanley," his mother said. "He means well. I think."

"We'll see!" Einstein called over his shoulder.

"And text me if you leave the neighborhood!" Mrs. Anderson shouted.

As he went out the back door, he took out his phone and texted Paloma again.

"On way to S — CU?"

The answer came back right away.

"SYS!"

Then Einstein got his bike out of the garage and rode down the driveway. He lived in a section of Sparta that had a mixture of private homes and low-rise apartment houses. The town was just between a suburb and the countryside. If you went out toward the highway, there were strip malls and fast-food restaurants. If you went a few miles in the other direction, you would soon hit farmland. In fact, Einstein's dad sometimes tended to sick farm animals.

To get to Stanley's house, Einstein had to

ride across Brookdale Park near the center of Sparta. From the top of the hill it looked like a picture postcard of some old time country town with a town square, a church with a white steeple, and a busy Main Street.

It was a beautiful day. The early-morning traffic had let up and few cars passed by. The sound of a power-driven lawn mower in the distance mixed with the sound of insects. The smell of newly cut grass hung in the air.

As he rode through the park, Einstein stopped to look at some red wood ants going into and out of an anthill. He took out his phone and took a photo for his science journal. Then he opened an app that identified insects and found out their scientific name, *Formica rufa*. He copied that into his electronic notebook on the phone.

His phone buzzed. It was another text from

Paloma.

"**OMW,**" it said. Paloma usually took her time getting anywhere, so Einstein knew he didn't have to hurry to meet her at Stanley's. He stopped to watch a downy woodpecker tapping for insects along a branch of a sugar maple tree; he threw some small granite rocks into the pond to test the strength of his pitching arm; and he observed that the cumulus clouds in the blue sky meant a fair-weather day. Finally, he got back on his bike and pedaled toward Stanley's.

Stanley lived in a complex of garden apartments and attached townhouses, new brick buildings clustered around green courtyards. At the entrance to Stanley's court, he saw Paloma waiting next to her bike.

Paloma Fuentes was Einstein's best friend. Some kids thought it was weird that his best friend was a girl, but that didn't bother Einstein. He and Paloma got along because they both loved science—except when they didn't get along because they disagreed about a case. Einstein would never admit it, but sometimes Paloma knew more about science than he did.

Paloma pushed the thick black hair out of her eyes and waved to him as he rode up. She was a little taller than Einstein, and was wearing old jeans, red high tops, and a white shirt that said "Geek Power" on the front. On her back was a small backpack. Einstein knew that inside the pack were a magnifying glass, some plastic specimen bottles, a guide to birds and wildlife, and other equipment. Paloma liked to be prepared.

"Hey, *Einstein!*" she called to him. She liked to tease him about his nickname.

Together, they walked their bikes to Stanley's

building and Einstein rang the bell. A minute later Stanley appeared at the door, carrying a large pair of normal-looking Rollerblades. He wasn't exactly glad to see them.

"Well, I guess I should have expected a visit from the science geeks," he said, as he stepped outside.

Stanley was a tall, thin kid with short, blond hair and a narrow face. Even though it was summer vacation, he was dressed very nicely in tan chinos, brown shoes, and a button-down shirt. Stanley liked to say he was always ready to make a deal.

"We were curious about your extreme Rollerblades," Paloma told him. "Is that them?"

Stanley hesitated, then he held the Rollerblades out, hanging from their laces. "That's right," he said proudly. "These are going to revolutionize skating. Anyone who buys a pair will go farther and faster than anyone ever has. Because ..." He paused dramatically. " ... these skates have no friction."

"Yeah, right!" Paloma laughed.

"It's true!" Stanley insisted.

"That's impossible," Paloma replied. "No one has ever built a machine with no friction."

"No one till now," Stanley replied smugly. "When the word gets out, everyone is going to want a pair of these. I'm going to sell them to kids all over town. And we can use the same technology for bikes and even cars. I'll be famous—and a billionaire."

"You'll be famous as a nut," Paloma said. She laughed. "Tell him, Einstein."

But as usual, Einstein didn't say anything— he just peered through his glasses at the skates, trying to get a better look.

"Okay," Paloma agreed, when Einstein didn't

reply. "I'll play along. How did you get rid of friction?"

"Hah!" Stanley laughed. "You think I'm going to tell you my secret?"

"Secret?" Paloma said. "The secret is how you can call yourself a scientist!"

Einstein interrupted the argument. "Do you mind?" he asked Stanley. Without waiting for an answer, he reached out and with the palm of his hand he spun the wheels on the nearest skate. They spun for a long time without making a sound and then slowly stopped.

"See?" Stanley said triumphantly. "See how they spin? My secret is going to make me rich and famous."

"I'm afraid not, Stan," Einstein said. "Besides, I already know your secret."

"Not possible!" Stanley cried. "Have you been spying on me?"

"No," Einstein answered. "I didn't need to."

"Your Rollerblades are terrific, Stanley," Einstein answered. He pushed back his glasses, which were slipping off the end of his nose. "But I'm afraid they are not friction-less."

"Come on, Einstein," said Stanley. "How can you know that without even trying them on?"

Can you solve the mystery? How did Einstein know that Stanley had not invented friction-free skates?

"Look at the wheels," said Einstein. "They stopped spinning."

"So what?" asked Stanley.

"But that's just it," said Einstein. "If the wheels were completely without friction, they would keep spinning forever. These wheels are very well greased, and they spin for a long time, but with even the tiniest amount of friction they will finally slow down and stop."

"You mean, it doesn't work?" said Stanley. He seemed genuinely surprised.

"They work," answered Einstein. "They just aren't friction-free. It's impossible to have two things touch each other with absolutely no friction. But you must have added some sort of lubricant to reduce the friction. I'd guess some kind of silicone spray."

"Yeah, that was it," said Stanley dejectedly.

"I knew it!" Paloma cried. "You can't get rid of all friction."

"Hey, don't feel bad, Stanley," Einstein said kindly. "Maybe your next idea is going to be a billion dollar one."

"It better be," said Stanley. "I have to pay back the money I borrowed for these skates."

"Well, look on the bright side—these skates are still revolutionary," Einstein said.

"They are?" Stanley said excitedly.

"They are?" Paloma echoed him.

"Sure," Einstein replied. "They have wheels don't they? And everyone knows when wheels go round they make revolutions."

From: Einstein Anderson
To: Science Geeks
Experiment: Build a Diving Submarine

Hey, fellow science geeks, it's me, Einstein Anderson. Want to make that diving submarine that made my kindergarten teacher call me "Einstein?" It's easy to do if you have the right equipment.

Here's what you need:

- A tall jar
- A balloon that's big enough to stretch over the mouth of the jar
- A medicine dropper
- Water
- A rubber band
- A pair of scissors

Fill the jar halfway with water. Place the tip of the medicine dropper into the water and squeeze the bulb to draw some water up into it. Then release the dropper in the water with the bulb up, so it floats with the top of the bulb just breaking the surface. Next, fill the jar with water to one inch from the top. If the dropper sinks, squeeze out some water. You'll have to get the right combination of water and air in the dropper so that it's light enough to float near the surface.

Now, cut out a circle of rubber from the balloon and fit it over the mouth of the jar. Secure the rubber tightly in place with the rubber band.

Press down on the rubber covering. The medicine dropper will sink to the bottom of the jar. Release the rubber covering. The dropper will rise to the top! The harder you press, the faster the dropper will fall. If you want a real challenge, see if you can press just hard enough to keep the dropper suspended midway in the jar. This is hard to do and requires constant pressure adjustments.

Check this out: What happens to the water in the dropper when you press down on the rubber? What happens when you release the rubber? What do you think is happening? Why does the dropper sink?

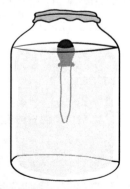

The Science Solution

When you press down on the rubber covering, air in the jar is compressed, which means that the same amount of air is squashed into a smaller space. Air can be compressed, but water can't, so when you press the rubber, some water is forced into the medicine dropper, which becomes heavier and sinks. When you release the rubber, the air in the dropper expands and pushes some of the water out, so the dropper becomes lighter and it rises.

This is also the principle that makes real submarines work! Submarines carry tanks of compressed air with them. They sink by taking in water and making themselves heavier. When they want to rise toward the surface, they use the compressed air to blow the water out, making the submarine lighter, so it floats.

So your medicine dropper really is a submarine. Anchors aweigh!

The Impossible Shrinking Machine

It was a Saturday morning, a few days after the mystery of the phony friction-free skates. Einstein and Dennis were in the backyard, playing catch. It was hot, the sun was shining, and Einstein was wishing they could go to the lake and go swimming.

"Hey, the sun is in my eyes!" Dennis squinted as he missed a catch. "Let's trade places."

"Okay," Einstein agreed. As they switched, Dennis looked puzzled.

"How come the sun wasn't in my eyes when we were playing yesterday?"

"Because yesterday we were playing in the afternoon," Einstein explained. "You know the sun rises in the east and sets in the west. It travels across the sky during the day."

"Sure, I knew that," Dennis said.

"Of course the sun isn't really going around the earth," Einstein continued. "The earth is, rotating, so it only looks like the sun is moving."

"Yeah, I knew that too," Dennis said, but he didn't sound so positive.

"Hey," Einstein asked as he picked up the ball. "Why did the teacher have to wear sunglasses?"

"Oh, brother!" Dennis groaned. "Not

another one!"

"Because her students were all so bright!"

They began throwing the ball again. Now with the sun in his face, Einstein really wished he could go to the lake. As if in answer to his wish, his mother came out into the backyard, holding a phone.

"It's for you, Einstein," she said. "Paloma has invited you up to the lake."

"Hey, *Einstein!*" she said when he answered the phone. Paloma always said "Einstein" like she was teasing him. "I'm at my Aunt Camilla's house, by Carter Lake. She said you could come up for the day tomorrow, if you want. Aunt Camilla's a biology professor at State University and has all kinds of science stuff at her house that you might like to see. Plus we can go swimming. Do you think your folks could drive you up?"

Einstein wasn't sure his parents could drive

him. But before he could say no, Paloma added, "Also, I've made a discovery up here that I want to show you. A new invention that's better than anything Stanley could dream up. I want to see if you can figure out how it works."

Well, that changed everything. Einstein couldn't turn down a science dare from Paloma. Out of all the science challenges he solved, hers were usually the hardest. He hung up and went inside to ask his folks about getting a ride.

His parents, Matt and Emily Anderson, were in the kitchen.

"I have to go out to the Bauerly place," his dad said. "One of their horses is sick. That's not far from Carter Lake, so I can give you a ride, if Paloma's aunt can bring you back."

Einstein's dad woke him up early the next morning and after a hearty breakfast, they drove up to Carter Lake. It was still early morning when they arrived.

Aunt Camilla lived in an old farmhouse, with a few sheds and small buildings around the property. Einstein had texted Paloma when they got close and now she was waiting outside. She was wearing a pair of overalls and a straw hat that kind of made her look like a farmer.

"Hey, *Einstein*!" she said, teasingly, as they drove up.

"Hey, Paloma," he answered as he got out of the car. "Why did the corn farmer have such good hearing?"

"I don't know," she answered with a groan. "Why?"

"Because he had so many ears!"

As usual, Einstein laughed at his own joke and Paloma just shook her head.

After Matt Anderson had said good-bye and driven off, Paloma turned to Einstein excitedly.

"Einstein," she said with a big grin. "I hope you're ready to be amazed."

"Sure," he said. "But that reminds me, why did the lab rat always look shocked?"

"That's easy," Paloma laughed. "Because he was stuck in *a maze*. You'll have to try harder than that, *Einstein*."

She led Einstein behind the farmhouse and down a twisting path into a grove of trees. Hidden from the house at the end of the path was a small shack with a bright yellow door. The early-morning sun shone directly on the yellow door and made it look almost like gold.

Paloma unlocked the yellow door and motioned Einstein inside. Einstein noticed that the single room they entered had no other doors and only one small window. The only objects in the room were a large stone table and a small black box sitting on the table.

"Here is my discovery, Einstein," Paloma told him. "I want you to look over the stone table closely. It was put together right in this room. You can see that it is too big to pass through the door or the window. You would have to break it into little pieces to get it out of the room."

Einstein checked the table carefully. He could see that what Paloma said was true. You would need a bulldozer to break up that old stone table.

"I'm now going to switch on my impossible shrinking machine," said Paloma. She flipped a switch on the side of the little black box. Nothing much happened except that the black box sort of burped once and then was quiet.

Paloma motioned Einstein to follow her out. "We'll have to leave the room so as not to shrink ourselves," she said. "But when we come back in a few hours, the table will be gone without a trace. The impossible shrinking machine will have reduced it down to the size of an atom."

"A shrinking machine?" Einstein repeated. "This is some complicated trick!"

"Oh, it's no trick, Einstein," Paloma said with a big grin. "At least, it's not one that *you'll* ever figure out."

Paloma led Einstein back to Aunt Camilla's house. For the rest of the morning Einstein and Paloma helped her aunt with a project she was working on. An invasive species of plant had gotten into Carter Lake. It was a kind of plant that wasn't native to this part of the world and so it had no natural enemies. No fish or other animals ate it. It was growing so fast it was choking off parts of the lake.

Since Aunt Camilla was a biologist she had

volunteered to help the town figure out a way to get rid of the plant. Einstein and Paloma helped by collecting plant samples from the edges of the lake and preparing them for Aunt Camilla to study under her microscope.

After lunch they went down to the beach and spent the afternoon swimming. By the time they got back to the house, Aunt Camilla had started an outdoor barbecue. They had grilled hamburgers, newly picked corn, a fresh tomato salad, and watermelon for dessert. It was all delicious, and they didn't finish cleaning up till seven o'clock. It was almost time for Einstein to go home.

"Well, Einstein," Paloma said with a grin. "Did you forget about my impossible shrinking machine?"

"No way!" Einstein exclaimed.

"Come on, then!" she told him.

The sun was going down, a blazing yellow disk over the treetops. Paloma led Einstein around the other side of the house and into the stand of trees from another direction. There was the shack, with the sun shining directly on the door, turning it golden, just as it had done in the morning.

Paloma unlocked the door, and they went inside. The room looked almost the same: one door, one small window, and one small black box. But the big stone table was gone. Nothing, not even a chip of stone, remained on the floor.

At first Einstein couldn't believe his eyes. How could that big stone table just disappear? Had Paloma really invented a shrinking machine? It didn't seem possible, but what was the other explanation? And where was the black box?

Paloma smiled at the look on Einstein's face. "Well," she asked, "what do you think of my impossible shrinking machine? I did it, didn't I? I finally stumped the great Einstein Anderson!"

She couldn't help herself and did a little dance of celebration.

Einstein was quiet for a few moments. Then his face changed, and he began to laugh. He pushed back his glasses, which had slipped down. "You almost had me there for a minute, Paloma," he said. "I think I know what happened to the table. And if I'm correct, there is no such thing as an impossible shrinking machine."

Can you solve the mystery? What do you think happened to the table?

"The key to solving this mystery," Einstein began, "is the sun."

"The sun!" Paloma exclaimed. "What does the sun have to do with the shrinking machine?"

"You know that the sun rises in the east in the morning and sets in the west in the evening," Einstein explained. "Yet both the rising sun and the setting sun shone directly on the yellow door. That's impossible."

"So what's the answer?" Paloma asked, but she already looked disappointed.

"Simple," Einstein said. "There are two rooms in the shack, one in back and one in front. And there must be two doors, on opposite ends of the building. The sun shone on one door in the morning and on the other door in the evening.

You must have taken me into one room in the morning and now we're standing in a different room. This room was always empty. The other room has the stone table."

"You're right," said Paloma with an air of defeat. "And for once I thought I had you beat! Come on."

She led him outside, and they walked around the shack to the identical yellow door. Paloma opened it and there was the stone table, just as it had been that morning.

"I see that I made one mistake," Paloma said, shaking her head.

"What's that?" Einstein asked.

"I should have shown you my impossible shrinking machine on a cloudy day."

"That's true," said Einstein. "Which reminds me—what do clouds wear under their shorts?"

"I don't know, Einstein," Paloma said with a grimace.

"Thunderpants!"

From: Einstein Anderson
To: Science Geeks
Experiment: Make a Sundial

Here's what you need:

- A pair of scissors
- A sharp pencil
- A map or globe with latitude marked on it—or access to the Internet
- A compass or analog watch

- A piece of heavy paper with a copy of this template printed on it—or glued on top.

Go to page 96 for a larger version of this template that you can copy and cut out for this experiment.

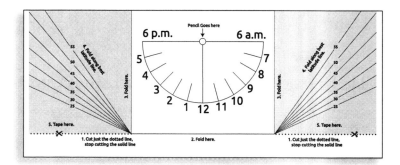

Or go to **seymoursimon.com/Einstein1** to print a copy.

1. Cut where it says "cut here" but stop when you get to the solid lines on each side.

2. Fold where it says "fold here" with the line on the outside. Crease, then open flat again.

3. Figure out the latitude where you are right now. You can find this by looking on a map or globe, or by checking on the Internet.

   Latitude lines indicate how far north or south you are from the equator. They are measured in degrees (°). For example, Austin, Texas is about 30° north, New York City is about 40° north, and the latitude of Moose Jaw, Saskatchewan, Canada is almost exactly 50°. I bet you will be the only kid in your class to know that!

4. Select the latitude line closest to your location, fold with the line outside, crease, then fold again with the line on the inside of the fold.

5. Tape the paper to a flat surface or piece of cardboard as shown here:

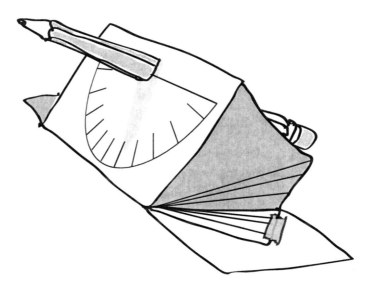

6. Insert a sharp pencil, point first, through the circle at the top-center of the template. Then take it out and re-insert it so the pencil points up.

7. Turn the sundial so that the pencil points due north. The best way to find north is to use a map or a compass. If you don't have a compass, you can use an analog watch or clock.

8. Hold your watch or clock up with the smaller hand pointing at the sun. Find the direction halfway between the little hand and the 12. That way is south and the exact opposite is north. (In case you're curious, you have to divide the distance in half because the hands go around a clock twice in a 24-hour day.)

Your sundial is done! The pencil's shadow will fall on the correct time of day. Not only that, the sun will be on the east side of the pencil *before* noon, and on the west side, *after* noon, which will remind us that the sun can't shine on the same door in Aunt Camilla's yard, morning and afternoon!

Why Does the Hound Dog Howl?

It was a quiet, sunny afternoon in the middle of July and Einstein was sitting under a tree in the backyard, his laptop in his lap, looking at the website for the Natural History Museum. He'd already been to soccer camp that morning and now it was just too hot to do anything but sit in the shade. His little brother

Dennis was home from day camp and was throwing a ball against the garage door with a regular thunk, thunk, thunk sound.

The sound stopped and even though he was absorbed in his reading, Einstein knew what to expect next.

"Hey, Einstein," Dennis said in a whiny voice. "Whatchya doing?"

"I'm making an apple pie," Einstein said, without looking up. "What's it look like I'm doing?"

"It looks like you're reading—again," his brother complained as he plopped down on the grass beside him.

"Oh, I guess I am," Einstein said, trying to sound surprised.

"Whatchya reading?"

Einstein looked up. Every now and then he had the idea of trying to teach his brother about science. It usually didn't work out, but he never stopped trying.

"It's an online exhibition about bees," he said, turning the screen around so Dennis could see the large color photos.

"Ugh!" Dennis said. "That's gross!"

"No, it's not," Einstein said, trying to stay patient. "Look at those compound eyes!" He enlarged the photo so Dennis could get a good look, but his brother pulled his head away.

"Get that away from me!" he exclaimed.

"It's just a photo," Einstein said, turning the screen back around. "Bees' eyes are really amazing. Did you know they can see ultraviolet light?"

"No. What's that?" Dennis sounded interested in spite of himself.

"It's a wavelength of light that we can't see," Einstein went on. "You know how when you hold a prism up to the light or you see a rainbow, you see different colors?"

"Yeah," Dennis replied.

"Well, those are the colors humans can see. But there are other types of light in wavelengths above and below the colors in a rainbow. We can't see them, but other animals can, including bees."

"Well, what good arc thcy, if we can't see them?" Now Dennis sounded annoyed.

"They may not be good for us, but animals use them to get around," Einstein replied. "Just like there are some sounds that we can't hear but other animals can. Like bats. They can hear high frequency sounds that humans can't. Isn't that amazing?"

"Uh, I guess," said Dennis. But he didn't sound amazed, just bored. And then he said with a whiny tone, "I'm bored. Let's do something."

Einstein sighed. The truth was, he was kind of bored, too.

"Hey!" he said. "What kind of bee is really bad at football?"

"Oh, no!" Dennis moaned.

"A fumble bee!" Einstein said, and laughed at his own joke.

Dennis got up and was about to leave in disgust, but when he turned he saw a tall boy standing in the driveway, wearing running shorts and a soccer team jersey. He held a big brown dog on a leash.

"Hey, Einstein. Hey, Dennis," the boy said, looking a little embarrassed.

"Hi, Pat," Einstein said. He was surprised to see Pat Hong standing in his backyard. They were in the same grade, but they weren't really friends. They didn't really have much in common. Even at the age of twelve, anyone could see that Pat was a natural athlete. He was good at almost every sport, but especially good at soccer.

49

While Einstein played in a league with kids his age, Pat was already playing on the town team with the high school kids.

Pat was good at sports, but he wasn't really interested in science. The only time he had ever talked to Einstein before was when he needed help with a science project for school.

"Hi, Einstein," Pat said again, looking a little awkward. His black hair was cut short, almost as if he had shaved his head. "I'm sorry to come over like this. I need to see your dad. He's a vet, right?"

Einstein got up and walked over to Pat and the dog. Dennis was already there, patting the dog's head.

"Yeah, he's a vet," Einstein said. "But he's not home now. Anyway, he sees patients at his office, not here."

"Oh!" Pat replied. "I guess I should have

figured that. But I don't have any way to get Rocky to his office. My mom's at work with the car."

Einstein bent over to look at Rocky. He was a large, mixed-breed dog who looked like he might be part poodle and part golden retriever. As Dennis petted him, Rocky wagged his tail happily.

"What's wrong with Rocky?" Einstein asked. "He looks pretty healthy."

"That's just it," Pat explained. "He seems fine but every now and then he starts howling for no reason. I'm worried he's in pain."

As if on cue, Rocky tilted back his head and gave a loud, mournful howl that lasted a few seconds. Dennis and Einstein jumped back in surprise.

"Wow!" Dennis shouted. "That dog really is sick."

"Maybe," Einstein said. He looked at Rocky,

who was now happily wagging his tail again. "But he seems better already."

"Is that dog howling again?" someone said from the end of the driveway. Standing on the sidewalk next to his bicycle was Stanley Roberts. He started walking toward them with a worried look on his face.

"Yeah, he's still sick," Pat said to Stanley. "He just starts howling for no reason."

"What do you mean, still sick?" Einstein asked, looking from Pat to Stanley. "Did you talk to Stanley about Rocky before this?"

"I offered him my advice," Stanley said, trying to sound very grown-up.

"And what advice was that?" Einstein asked.

"Stanley wanted me to buy these dog vitamins he invented," Pat told him.

"Vitamins?"

"Dr. Roberts's Dog Formula, from StanTastic Industries," Stanley said proudly, producing a large brown jar with a homemade label on it. "Guaranteed to make your dog healthy and happy. These are just the first batch."

Einstein held out his hand and, reluctantly, Stanley gave him the jar. Einstein unscrewed the cap, looked inside, smelled the contents, then screwed the cap back on and handed the jar to Stanley.

"I tell you what, Pat," Einstein said, taking his phone out of his pocket. "I'll call my dad and we'll ask him what we should do. How does that sound?"

"Great," Pat said, looking relieved.

"Okay," Stanley told them. "But the cure is right here and I can let you have it for a special discount, today only."

"Uh, no thanks, Stanley," Pat told him.

Stanley shrugged and walked back down the

driveway. Einstein watched him as he got his bike and walked it down the street. Just before he disappeared behind the house he saw Stanley take something shiny out of his pocket. A second later Rocky began to howl again. This time the howling lasted for a few seconds. Dennis put his hands over his ears.

"Make him stop!" Dennis cried.

"Yeah, Einstein," Pat pleaded. "Can't you make him stop?"

Stanley came running back up the driveway.

"I heard Rocky all the way down the block," he said. "I think you'd better buy these vitamins, Pat. You don't want him to get any worse, do you? If the neighbors complain you might have to give him away."

"No, I don't want that," Pat said, looking very worried. "Okay, how much are they?"

"Hold on, Pat!" Einstein said. "I don't think you need those so-called vitamins at all. I know why Rocky is howling and I know how to make him stop."

Can you solve the mystery? Why was Rocky howling?

"**Oh, so now you** think you can cure sick dogs?" Stanley sneered.

"I'm no more a veterinarian than you are, Dr. Roberts. But Rocky isn't sick."

"He's not?" Pat asked.

"No, the problem with Rocky is in Stanley's pocket."

"My pocket?" Stanley blustered. "You're crazy!"

Einstein turned to Pat. "Let me ask you—did Rocky ever howl when Stanley wasn't nearby?"

"Uh, no, now that you mention it. Each time he howled, Stanley either showed up right after or had just left."

"That's what I figured," Einstein said. "Stanley has a dog whistle in his pocket."

"A dog whistle?" Dennis asked. "Why do dogs need a whistle?"

"It's not a whistle for dogs to use," Einstein explained. "It's a whistle that only dogs can hear. A dog whistle is so high-pitched that people can't hear the sound, but dogs can hear it even from a distance."

"Is that true?" Pat said angrily, taking a step toward Stanley. He towered over him.

"Uh, look, I was just trying to get my vitamin business started," Stanley said nervously. Out of his pocket he took a shiny, round whistle.

"I can't believe you hurt my dog to sell your stupid vitamins!" Pat shouted.

"It didn't hurt him," Einstein said quickly, before Pat could really lose his temper. "Rocky just heard the noise and answered it. And those vitamins look and smell like they're regular old dog treats. Isn't that right, Stanley?"

"Yes," Stanley said unhappily. "You got me again, Einstein. But one day, you won't be so smart." He turned and stomped off down the driveway.

"Gee, thanks, Einstein," Pat said. "Don't you worry about Stanley. I've got your back."

"Thanks, Pat, but Stanley is pretty harmless. Just sometimes his greed gets the better of him. Hey, do you know what is a dog's favorite vegetable?"

"No," Pat replied.

"Get ready for a stinker," Dennis warned with a groan.

Einstein had a big grin on his face. "Come on, Pat, everyone knows a dog's favorite vegetable is a *collie* flower!"

From: Einstein Anderson
To: Science Geeks
Experiment: Start a Bottle Band

Here is a puzzle for you, fellow scientists. Can you figure out how to make a bottle filled with water make both a high-pitched sound and a low-pitched sound?

Here's what you need:

- 3 bottles—remove any labels
- Water
- A drumstick or spoon

Fill a bottle halfway with water and blow across the top until you make a good sound. Fill the other two bottles: one more than half full and the other less than half full. Now blow across the top and listen to the sound. What happens? Bottles with more water emit a higher-pitched sound and bottles with less water make a lower- pitched sound.

Now—once you have found some sounds you like, see if you can make the same bottles, with the same amounts of water, make a completely different sound. This time, the bottle with more water will make a low sound and the one with less water will make a higher sound. How can this be?

Did you figure it out? You're a smart kid and you know I wouldn't say you need something if you don't need it. So what's the drumstick or spoon for? You got it! If you hit the bottle with a drumstick, a full bottle makes a low sound and one with less water makes a higher sound. Can you figure out why?

The Science Solution:

That's it! When you blow across the top of a bottle, the air inside vibrates. If there's more air, it makes a lower sound and if there's less air (and more water), it makes a higher sound. When you hit the bottle with a spoon, it's the water and the bottle that vibrate, not the air. A lot of water vibrating makes a low sound and a little makes a higher sound.

These are the basic principles behind a lot of musical instruments. Maybe you and a friend or two could make a band, using instruments you make from bottles and other discarded things. Check out this amazing group in Northern California. They're members of the San Francisco Youth Orchestra playing a piece of music composed for the "junkestra."

Go to **seymoursimon.com/Einstein1** for a link to see the Junkestra orchestra play.

Can you figure out what "junk" is making each sound?

The No
Treasure Hunt

"That's Megaptera novaeanglia.
N–O–V–A–E–A–N–G–L–I–A."

Paloma Fuentes was carefully reading the words from her laptop screen aloud as Einstein typed in the letters on his.

"*Megaptera?*" said Einstein's mother. "That's the name for a kind of whale, isn't it?" She was

62

seated just across from them at the kitchen table in Einstein's house. Emily Anderson was working from home that afternoon, writing a story for the *Sparta Tribune* on *her* laptop.

"Yes, that's right, Mrs. Anderson," Paloma replied, sounding a little surprised. "It's the scientific name for humpback whales. *Megaptera* means great wings. Of course the humpback doesn't have wings, but they have long flippers that look sort of like wings. *Novaeanglia* means New England. That's where the humpback whale was first seen by Europeans. We're doing some research on whale sonar."

"Oh, I see. Thank you for the explanation, Paloma," Mrs. Anderson told her with a little smile. As Einstein Anderson's mom, she was used to getting a lot of scientific information in a single sentence.

Einstein's brother Dennis walked through

the kitchen on his way to the backyard.

"Who wants to study fish in the middle of vacation?" he asked as he looked over Einstein's shoulder at the pictures on the screen.

"These are whales, not fish," Einstein explained, trying once more to get Dennis to pay attention to science. "Whales are mammals, and like all mammals, they breathe air, have hair or fur, and mothers feed their young with milk. Also, almost all mammals give birth to live young."

Dennis gave Einstein a blank look.

"That means they don't lay eggs—except for the platypus and a few other species in Australia."

"Well, then you must not be a mammal, Einstein," Dennis said. "Because you just laid an egg by doing schoolwork during the summer."

Across the table, his mom chuckled.

"Well, I guess you're not the only one to make bad puns, Einstein," she said.

Einstein couldn't help laughing himself. "I guess so."

"You know, you kids might be interested in this story I'm doing." Mrs. Anderson pointed to her laptop screen. "It's about a new store that's opening on Main Street called Things of Nature. They sell all sorts of amazing objects. Here, I'll send you the link to their website."

She moved her fingers over her keyboard and a few seconds later, Einstein had an email with the link to the store's website. He then sent it to Paloma.

"Look at this, Einstein, they have a contest!" she said, as soon as the site opened on her screen.

"I'm way ahead of you," Einstein replied.

"Oh, yes, I thought you two would like that," his mom said. "They'll give a fifty-dollar gift certificate to anyone who can spot a fake object in their store. I'll take you down there to check it out if you want. I have to go anyway for the story."

Einstein and Paloma didn't answer, however. They were too busy studying the objects displayed on the store's website.

"Look at those quartz crystals," Paloma said a few moments later. "And those mounted insects. And those animal skeletons."

Einstein was looking at the same photos. The store certainly did sell a lot of incredible stuff. There were all sorts of minerals and semi-precious stones, and hunks of pyrite, or fool's gold. There were even rocks that the store said were real meteorites. There were real stuffed animals, like a real alligator head and a large lizard called a bearded dragon.

"Now, those are cool!" Dennis cried looking at the reptiles over Einstein's shoulder.

"Einstein, when you win the gift certificate, buy one of those!"

"We'll see," Einstein said. Then he clicked on another tab and looked at fossils the store was selling. Those included petrified wood and something that looked like a bug they called a trilobite.

"Petrified wood?" Dennis asked. "What's it scared of?"

"It's not that kind of petrified," Paloma told him. "That means it's wood that has turned into a rock. That rock used to be a part of a tree."

"That sounds like a fake to me," Dennis said, shaking his head. "Whoever heard of a tree that became a rock?"

The store sold all sorts of other items. On one page were photos of eggs they said were from the golden eagle, the turkey vulture, the brown bat, and the American condor. Then

there were seashells, including tritons, cowries, and one that was labeled, "giant frog shell." Finally, Einstein looked at the pages of insects, multicolored beetles, and butterflies.

"I think we should go down there, Einstein," Paloma said excitedly, as she looked up from her laptop screen. "I bet if we looked around we could find at least one fake. What about these meteorites? I bet they're not real."

"Maybe," Einstein said. "But real meteorites aren't that rare. It's possible they could have some for sale."

"What about these bugs, then?" Paloma asked. "Some of them look like they've been painted."

"Maybe," Einstein agreed. "But it's easy enough to check those out by looking online."

"Hey, what about that giant frog shell!" Dennis cried out, pointing to the photo. "Frogs don't have shells!" He paused, then added, "Do they?"

"Oh, I know that one," Paloma told him. "That's just the name of the shell. It's not from a real frog." She turned to Einstein. "Come on, *Einstein*," she teased him. "Let's go win the prize!"

"Yes, Einstein," his mother said with a smile. "I never thought I'd see you back down from a science challenge. And here I was all ready to drive you kids down there so you could find the fake object and win the gift certificate."

"Oh, I'm not backing down," Einstein replied. "It's just that there's no need for us to go to the store. I've already spotted the fake."

Can you solve the mystery? Which object is a fake?

"You have?" his mother said, sounding impressed. "Don't you have to see the objects in person?"

"Not in this case," he told her.

"It's the tree rock, isn't it?" Dennis cried. "I was right. It's the petrified wood. That's the fake."

"I'm afraid not," said Einstein. "Petrified wood is real. It's formed when the wood cells in an old piece of wood are replaced by a chemical called silica. Silica turns into a rock. The process takes thousands of years, but it's real."

"Well, if it's not the rocks, then it must be the bearded dragon," Dennis said. "Who ever heard of a dragon with a beard?"

"I have," Einstein said. "It's a kind of reptile that lives in Australia."

"Then which one's a fake?" Dennis asked.

"The one object that is surely a fake is the brown bat's eggshell," Einstein said.

Paloma smacked her hand on her forehead.

"Of course!" she cried. "How could I have missed that?"

"How do you know?" Dennis asked.

"Do you want to tell him, Paloma?" Einstein said.

Paloma nodded. "Because bats are flying mammals, not birds," she said. "Their wings are really arms with long fingers. A thin, furry skin stretched between the fingers forms the wing. And of course bats don't lay eggs like birds. Bats, like most other mammals, give birth to living young."

"I'm impressed!" Einstein's mother said. "I guess you didn't have to visit the store, after all."

"Oh, yes we do," said Einstein. "We have to go collect our prize! But we have to pay attention when you drive us, Mom. We don't want to miss our *egg-sit!*"

"Oh!" Dennis groaned. "But that reminds me—I'm hungry."

From: Einstein Anderson
To: Science Geeks
Experiment: Discover the Warmest Coat

What I like about science (aside from the fact that it's fascinating and full of surprising, fun experiences) is that you can use scientific principles to solve real-world problems. Take, for example, the birds and mammals that Paloma, Dennis, and I were looking at on the Internet. Birds and mammals are different in a lot of ways, but they both face some of the same problems. They've just come up with different ways of solving them.

Think about this: It's the middle of February. Snow covers the ground and it's so cold that a thin, crunchy crust of ice lies on top of the snow. Freezing wind moans through the bare branches . . .

Feeling chilly yet?

Now let's say there are three creatures out in that storm—a fox, a crow, and a person. How will they keep warm? The fox has a nice, thick coat of fur and the crow has feathers. Luckily, the person has a warm coat. But wait a minute. Touch the outside of the coat. That coat is not warm—in fact it's freezing cold on the outside. How does the cold coat keep the person warm?

Let's do an experiment and figure it out!

Here's what you need:
- 3 jars with lids
- Hot water from the sink (ask an adult for help)
- A thermometer
- A piece of wool (a sock or a sweater will do)
- Some feathers or a down jacket or blanket—use a whole down-filled jacket sleeve or blanket. Don't cut anything!
- A piece of paper

1. Fill up the jars with hot water and quickly wrap them—one in wool, one in feathers, one in paper. Use the thermometer to record the temperature in each jar before you screw on the lid.

2. Let them sit for 10 minutes.
   Check the temperature again in each jar.

3. Check once more in another 10 minutes.

What did you notice? Which water is the warmest? Which water is the coldest? Why did the wool and the feathers keep the water warmer longer than the paper?

The Science Solution:

It's not that wool is hot—it's that it kept the heat from escaping from the jar. It created insulation, to trap the heat close to the jar and keep it warm. In fact, what makes wool warm is that it has a lot of dead air between the fibers! Feathers work the same way. The feathers have tiny filaments that are intertwined, trapping a lot of dead air between the bird's warm body and the outer feathers, which act as a barrier between the warm bird and the cold air. Just like the person's coat, the outside of a bird's feathers or a fox's fur may feel cold to the touch, but the air spaces between the hairs or the feathers are keeping warmth trapped inside. Air is a much better insulator than paper.

Pretty cool, don't you think?

Solving the Dissolving

"StanTastic Industries' latest super product! Super solvent dissolves anything, anywhere, anytime. Will revolutionize industry!"

Paloma read out loud the words on the small screen of her phone. Then she turned to Einstein, who was straddling his bike in the driveway of his house.

"Can you believe that guy?" she asked. "He just never stops with his crazy schemes."

"Well, at least he's a hard worker," Einstein said calmly.

"Hard at work trying to become the first twelve-year-old billionaire," Paloma laughed. She'd ridden her bike over as soon as she'd seen the new advertisement on Stanley's blog. "Do you think we should check it out?"

Einstein hesitated. "I don't know," he said slowly. "Mr. Green probably won't let him get into any trouble."

Mr. Green was the chemistry teacher at Sparta High School. He ran a science club for kids during the summer. That's where Stanley usually made his "inventions." Paloma and Einstein were members of the club, too.

"Mr. Green doesn't always pay attention to Stanley," Paloma pointed out. "Remember the time he almost blew up the lab?"

"That's true," Einstein replied with a sigh. "Well, I guess we'd better get over there. I'll call my mom."

Einstein took out his phone and tapped the screen to call his mom, who was at work at the *Sparta Tribune*. It was a Wednesday afternoon. Soccer camp was over for the day, Dennis was at a friend's house, and his dad was at his veterinarian's office.

"That's good," his mom said, a little absently, when he told her where they were going. "Try not to be too hard on Stanley. He means well."

Einstein hung up and he and Paloma started pedaling toward the high school, only a few blocks away. Even though they were just entering middle school, they had been there many times for the science club. They locked their bikes to the bike rack and ran up the broad front steps, then up to the second floor lab.

As they suspected, Mr. Green wasn't paying any attention to Stanley. He was absorbed with a couple of high school kids who were building a robot for a regional robotics competition. He waved to Einstein and Paloma as they walked in, then turned right back to the robot.

Stanley was by himself at the other end of the room, sitting in front of a counter jammed with flasks, tubes, and other chemistry equipment. Some of the flasks were filled with red, green, or blue liquids. A maze of glass and plastic tubing connected the beakers one to another.

Einstein had no idea what was in the different containers and he suspected Stanley didn't either. But it *did* look very scientific.

Stanley looked up as the two of them approached.

"Hmm," he frowned. "I was wondering when you two snoops would show up. Come to steal my big invention?"

"Just came to see what you were up to," Einstein said, trying to sound friendly.

"Eww!" Paloma said, holding her nose. "What's that smell?"

"It's a perfume formula I'm working on," Stanley said proudly. "All celebrities have perfumes these days. When I'm a world-famous billionaire, I'll have one, too."

"You'll need one," Paloma muttered, and Einstein nudged her with his elbow.

"But that's just a sideline," Stanley agreed. "I've been too busy with my new invention. It's a sure thing." He paused, then trying to sound very grown-up, he added, "In fact, it's good you're here, because now you have a chance to be the first to invest."

"Invest?" Einstein repeated. Then he gave a little chuckle. "That's what my mom says when she wants my dad to put on a three-piece suit. In *vest!* Get it?"

Paloma groaned, but Stanley just ignored him. Instead he pointed to a maze of tubing. At the end of it sat a small beaker half-filled with a clear red liquid. "That's it," he said proudly. As he spoke, another drop of red liquid plopped into the beaker.

"What is it?" Einstein looked at the beaker curiously. "That looks like a glass of cherry soda pop to me."

"Hah!" Stanley laughed in his best imitation of a mad scientist. "It looks like soda pop," he declared, rubbing his hands together gleefully, "but that liquid is the first universal solvent ever made."

"Really?" Einstein said. "If that's true, you really might become a billionaire."

"Of course it's true," Stanley insisted.

"Stanley, a solvent is just a liquid that dissolves another substance," Paloma pointed out. "Water is an excellent solvent. It dissolves sugar, salt, and many other things. In fact, give it enough time, and it can dissolve almost anything. That's why some people even call water the universal solvent. Is that just colored water in the beaker?"

"No!" Stanley insisted angrily. "You think I called you over here to show you some colored water? Besides, water doesn't dissolve everything. For example, water doesn't dissolve oil or fingernail polish. But you can use turpentine to dissolve oil, and nail-polish remover to dissolve nail polish. There are lots of different solvents for different things. But my solvent dissolves anything."

"All right," Einstein said. "Now I'm curious. Let's see your solvent work."

"What would you like me to dissolve?" Stanley asked.

Einstein looked around the laboratory. "Try a piece of chalk," he said, "and some linseed oil, and some borax." Einstein knew that the chemicals he had chosen did not dissolve very well in water.

"Sure," Stanley said. He set up three more beakers with the red liquid. He dropped the chalk into one beaker and each of the other chemicals into the other two beakers. This time it took somewhat longer for anything to happen. But sure enough, the chalk and the chemicals slowly began to dissolve.

Einstein was impressed. He looked closely at the beaker.

"Well, that's not water," he said. Paloma nodded in agreement. She took out her phone.

"Okay, it's not water," she said, quickly

searching the Internet. "What is it? Acetone? That's a pretty good solvent. Look."

She held up the phone's screen so Stanley could see the page she'd found about acetone.

"I'm not saying," Stanley said. "That's the billion-dollar secret!"

"Okay," Paloma nodded, putting her phone away. "What else can it dissolve?"

Stanley took a piece of plastic and placed it in the beaker. He stirred it around with a glass rod. It got smaller and smaller. After a few minutes the piece of plastic had dissolved.

Stanley looked over at Einstein and said, "See, I told you—it's a universal solvent. Even the great Einstein Anderson has to admit I did it!"

Einstein looked thoughtful. He picked up each beaker of red liquid and looked at it carefully. Then he pushed his glasses back onto his nose and smiled. Paloma looked at him curiously. Had Stanley finally stumped them?

"Stanley," Einstein said, "I wish you really would invent something super, but I don't think this is it. That red liquid you made may dissolve lots of things, but I'm sure it's not a universal solvent."

Can you solve the mystery? How did Einstein know that Stanley had not invented a universal solvent?

"If that was a universal solvent," Einstein declared, "you'd be in a lot of trouble."

"What kind of trouble?" asked Stanley. Suddenly his expression changed from very confident to very worried. "You mean with the government?"

"Not the government," Einstein answered. "More like the school ... or anyone else who tried to use it."

"What are you talking about, Einstein?"

"Just this," Einstein said. "A real universal solvent will dissolve anything, even glass. So if your cherry soda were a universal solvent it would dissolve the glass beaker and the stirring rod. No container could hold it. It would dissolve anything it touched, including the floor of this classroom."

Paloma laughed. "I didn't think of that, Einstein," she said.

Stanley slowly nodded. "I guess you're right, Einstein," he grumbled. "Again!"

Einstein shrugged his shoulders. "Don't feel bad, Stanley. Hey, when you start selling your perfume, do you know what you should use to deliver it to people?"

"No, what?" Stanley grumbled.

Einstein started laughing before he could get out the punch line. "A smelly-copter!"

"Oh no!" Paloma groaned. "That joke really stinks!"

From: Einstein Anderson
To: Science Geeks
Experiment: Make Rock Sugar Candy

Well, Stanley's bug juice may not have been a universal solvent, but it did dissolve a lot of things. Another fantastic solvent is something we use every day—water! If you've ever had a leak in your roof, you know that water can cause a lot of damage. But did you ever think about what happens to the stuff that dissolves in water? Can you get it back once it has dissolved?

Let's do a sweet experiment with water—and make some candy!

Here's what you need:

- 1 cup water
- 3 cups table sugar (sucrose)
- A clean glass jar
- A pencil
- String (cotton or wool works better than nylon)
- A pan for boiling water and making solution
- A spoon or stirring rod
- Food coloring (if you want to make your candy colorful)

Two warnings:

1. You need boiling water for this experiment, so you should ask an adult to help you.

2. You will have to wait for your candy—a few days or up to a week—but it will be worth it!

Boil the water and slowly pour in the sugar, stirring constantly. Watch the sugar dissolve in the water. Keep stirring and adding sugar until it starts to accumulate on the bottom of the pan and won't dissolve, no matter how much you stir. At this point, the sugar solution is "saturated."

This is the time to add food coloring, if you want it. Personally, I like clear candy, but red is pretty cool, too!

With adult help, pour the sugar solution into the clean glass jar. If there is un-dissolved sugar at the bottom of the pan, don't let it get into the jar. Tie a piece of string to the middle of the pencil. If you want, you can tie a paper clip to the bottom of the string to keep it hanging down straight. Put the pencil across the top of the jar with the string hanging down inside. Don't let the string touch the bottom or sides of the jar.

This is the hard part—put the jar someplace quiet and don't disturb it!

Check it out the next day and the next. Do you begin to see something growing on the string? This might be a good experiment to do on Friday and leave it over the weekend, so you don't accidentally try to taste it before it's ready.

In a few days you'll have some tasty rock candy crystals. When they stop growing larger, take the string out of the water, let it dry, and crunch some sweet crystal sugar.

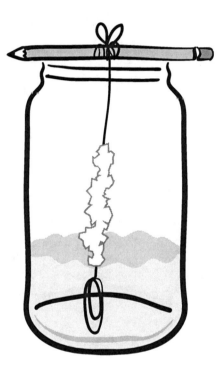

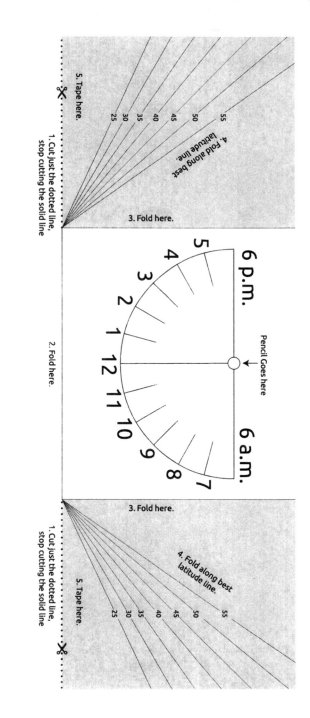

5. Tape here.

1. Cut just the dotted line,
stop cutting the solid line

4. Fold along best latitude line.

3. Fold here.

55 50 45 40 35 30 25

2. Fold here.

6 p.m.

Pencil Goes here

6 a.m.

5
4
3
2
1
12
11
10
9
8
7

3. Fold here.

1. Cut just the dotted line,
stop cutting the solid line

5. Tape here.

4. Fold along best
latitude line.

55 50 45 40 35 30 25

Go to seymoursimon.com/Einstein1 to print out a copy of this template.

The template in this book only works if you live in the Northern Hemisphere.
But you can find one for the Southern Hemisphere on the website.

CPSIA information can be obtained at www.ICGtesting.com
Printed in the USA
LVOW12s1517291113

363228LV00019B/1100/P